I0005016

Software Project Plans

A How To Guide for Project Staff

David Tuffley

To my beloved Nation of Four
Concordia Domi – Foris Pax

*Planning is bringing the future into the present
so that you can do something about it now.*

Acknowledgements

I am indebted to the Institute of Electrical and Electronics Engineers on whose work
I base this book, specifically IEEE Std 1058 Recommended Practice for Software
Design Descriptions.

I also acknowledge the *Turrbal* and *Jagera* indigenous peoples, on whose ancestral
land I write this book.

Contents

Contents

A. Introduction

The project plan documents the planning work necessary to conduct, track and report on the progress of a project. It contains a description of how the work will be performed.

The benefit of having this standard is the consistency of presentation, enabling management to assess the plans, for their merits or limitations, more readily.

In particular this standard specifies the format and content for a project plan by defining the minimal set of elements that shall appear in all project plans (additional sections may be appended as required).

The project plan includes the following.

- The scope and objectives of the project.
- The deliverables the project will produce.
- The process for producing those deliverables.
- The time frame and milestones for the production of the deliverables.
- The organisation and staffing which will be established.
- The responsibilities of those involved.
- The work steps to be undertaken.
- The budget.

A.1. Scope

This standard applies to Projects operating under the software quality management system.

This standard does not apply to small projects for which approval has been obtained to use a terms of reference in lieu of a project plan, quality plan and customer agreement.

A.2. Objectives

The objective of this standard is to provide project managers with a guide for the development of the project plan.

The outcome of using this standard will be the following.

- All specified aspects of the project will be considered during the project planning stage.

- Project plans will have consistent content and format.

- Projects will have plans that specify their objectives, deliverables and manner of execution.

A.3. References

IEEE 1058 - Standard for software project management plans

A.4. Definitions & acronyms

Work Package   This is the IEEE term for task of work.

Project   In this standard the project manager is
Manager   the person managing the project for the
                developer. In the project plan the name
                for this role may be changed as long as all
                terms are covered in the definitions and
                acronyms section of the plan.

A.5. Responsibilities

The Project Manager is responsible for the following.

- Preparation and maintenance of the project plan.

- Presentation of the project plan as part of the project approval process (Std-09).

- Conducting the project in accordance with the approved project plan.

B. Software Project Plan

An approved project plan is a prerequisite for a project. It must be created at the commencement of each project conducted under the software quality management system. The project plan is developed at the same time as the project agreement and the quality plan.

Note: Small projects for which approval has been obtained (refer Quality Manual) can use a terms of reference (Std-10) in lieu of a project plan, quality plan and customer agreement.

All project plans shall conform to Std-01, be filed in accordance with Std-03, and change controlled in accordance with Std-18.

This procedure description section outlines the minimum content of the project plan. It is derived from the IEEE Standard for Software Project Management Plans (IEEE 1058.1-1987). FM-06 is provided to assist in the development of project plans.

All project plans shall have the following contents, as described in the table of contents sections following.

Figure 1 - Project plan table of contents

```
Table of Contents
1.  Introduction
    1.1. Project objectives
    1.2. Project deliverables
    1.3. Evolution of the project
    1.4. Reference materials
    1.5. Definitions and acronyms
2. Project organisation
    2.1. Life cycle model
    2.2. Project organisational structure
    2.3. Project organisational boundaries and
interfaces
    2.4. Project responsibilities
        2.4.1. Project manager
        2.4.2. Steering committee
        2.4.3 Customer representative
3. Managerial process
    3.1. Management objectives and priorities
    3.2. Assumptions, dependencies and constraints
    3.3. Risk management
        3.3.1. Risk management plan
        3.3.2. Result of analysis
    3.4. Monitoring and controlling mechanisms

    3.5. Staffing plan
4. Technical process
    4.1. Methods, tools and techniques
        4.1.1. Methodology
        4.1.2. Techniques
        4.1.3. Tools
    4.2. Software documentation
    4.3. Project support functions
        4.3.1. Quality
        4.3.2. Verification and validation
        4.3.3. Configuration management
        4.3.4. Administration/other
5. Work packages, schedule and budget
    5.1. Work packages
    5.2. Dependencies
    5.3. Resource requirements
    5.4. Budget and resource allocation
    5.5. Schedule
6. Project training plan
7. Index (optional)
```

Where a section is not relevant, still include the section and say 'Not applicable' with a brief reason(s) for leaving it out.

Additional sections may be added as required and should follow a similar format and numbering convention as the rest of the project plan.

Some of the material may appear in other documents. If so, then reference to these documents shall be made in the body of the project plan.

B.1. Introduction (section 1)

Section 1 of the project plan provides an overview of the project.

B.1.1. Project objectives (section 1.1)

This section contains a concise statement of the objectives of the project. These normally relate to the project management aspects of the project. Do not restate the production of each deliverable as an objective, and do not restate the Quality Objectives as these are given in the Quality Plan.

A few examples of possible project objectives are given below.

1. Complete the project within +/- 20% of its approved budget.

2. Deliver x% of ad hoc customer requested intermediate deliverables by the date required.

3. Deliver all final deliverables as specified in the project agreement (Std-05) by the date required.

4. Maintain a minimum customer satisfaction rating in accordance with Std-22, Customer Feedback, of "Very satisfied" throughout the project.

In addition to stating these objectives, this section shall contain a table showing the method of verifying whether the objective has been met. This will typically refer to a metric or deliverable that can be used to determine whether the objective has been met. **For example:**

Verification that these objectives will be achieved shall be through the following.

Objectives	Verified by
1	Project financial report (in accordance with Std-23)
c2	Specially kept project metrics on number of ad hoc customer requested deliverables, date requested and date supplied. (refer Quality Plan, section 5.3)
3	Project financial report (in accordance with Std-23)

Table 1. Project verification for the objectives.

B.1.2. Project deliverables (section 1.2)

This section lists all project deliverables and shall reflect the deliverables as stated in the most up-to-date customer agreement either directly or by reference. Interim deliverables which are produced only to support the internal project process are not specified here.

For example:

The following are the deliverables for the project.

Item Description	Date
Requirements specification	dd/mm/yy- Delivered
Project Plan, Quality Plan and Schedule	dd/mm/yy - Delivered
Production of prototype	dd/mm/yy - Report Delivered
Production of user documentation	dd/mm/yy - Delivered

Table 2. Project deliverables.

B.1.3. Evolution of the project (section 1.3)

This section clarifies the status of the project at the time of creation of the project plan. Generally, the following should be stated.

- Summary of the business case from which the project gained approval.

- Approval status for the project.

- Summary of existing documentation created during the project initiation process.

Note that these items refer to the **project**, not the **product**.

In addition to the above the section describes the means by which scheduled and unscheduled updates are made to the project plan. This includes mention of how the changes are to be distributed.

B.1.4. Reference materials (section 1.4)

This section contains the list of all documents referenced in the project plan. The references shall be formatted in accordance with Std-01.

Any deviations from referenced standards shall be identified and justification given for the departure. This information shall be included at the place where the standard is referred to in the body of the project plan, not here in the references section.

B.1.5. Definitions & acronyms (section 1.5)

This section contains a list of the terms used in the project plan which may need special knowledge or interpretation. As a guide, words not occurring in standard dictionaries should be defined. Names such as Tuffley Computer Services are exceptions to the rule.

B.2. Project organisation (section 2)

Section 2 of the project plan defines the life cycle model and organisational issues relating to the project.

B.2.1. Life cycle model (section 2.1)

This section depicts the major phases or stages of the project. Many different project structures may exist given the major differences between, for example, a large development, purchase of a package, enhancement of an existing system and so on.

This model must show major milestones, with accompanying baselines, major review points and deliverables.

This section contains only a high level project phase description of the project's life cycle. It does not include the detailed tasks or activities necessary for the construction of the schedule. These are given in section 5.

An example of a life cycle model appears in Figure 2 following.

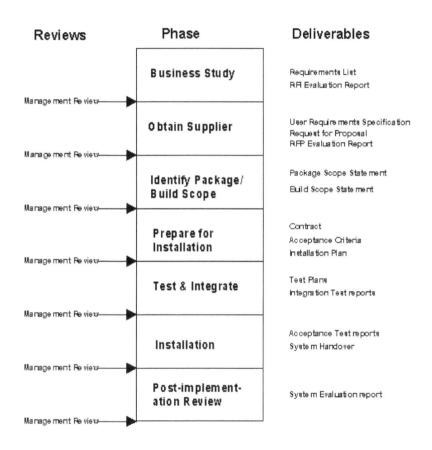

Figure 2 - Project Life cycle Model

B.2.2. Project organisational structure (section 2.2)

This section specifies the internal structure of the project team. It describes the lines of reporting and communication between team personnel from the Project Manager down.

An organisation chart would best depict the structure. Supporting text material would provide clarification where needed.

Note: The quality organisation, which extends beyond the project team, is detailed in the quality plan.

B.2.3. Project organisational boundaries & interfaces (section 2.3)

The purpose of this section is to describe the administrative and managerial interfaces within the project and between other relevant entities external to the project.

The following sample relationship diagram depicts typical interfaces where the Project Manager represents the project.

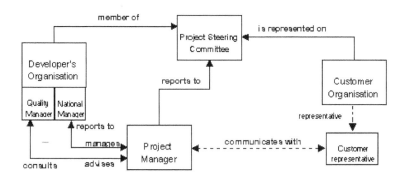

Figure 3 - Organisational Interfaces

B.2.4. Project responsibilities (section 2.4)

The purpose of this section is to identify the types of responsibility applicable to each major function in the project. Types of responsibility may be to perform, to review, to assist, to consult, or to sign off, so that different individuals can have responsibilities for the same project function. It is recommended that matrix be used to depict these responsibilities.

Two major functions that are a part of all projects are the following.

B.2.4.1. Project Manager (section 2.4.1.)

Responsibilities of the Project Manager are set out in the document 'Project Manager Terms of Reference', which will be produced by the Steering Committee during Project Initiation. Refer to Std-10 - Terms of Reference.

Note: In this standard the term project manager is used. In the project plan the name for this role may be changed as long as all terms are covered in the definitions and acronyms section of the plan.

B.2.4.2. Steering Committee (section 2.4.2.)

Responsibilities of the Steering Committee are set out in the document 'Steering Committee Terms of Reference'. Refer to the Terms of Reference document.

A third major function that may be part of a project is the following.

B.2.4.3. Customer Representative (section 2.4.3.)

Responsibilities of the customer representative are set out in the Project and Service Agreement (Std-20).

The following is an example of a responsibility matrix.

Respons-ibility Matrix	Responsible Organisational Element			
Respons-ibilities	Project Mgr.	Quality Manager	General Manager	Customer Rep
1. Project plan	P	R	A	R/Ac

2. Quality plan	P	R/A		R/Ac
3. Project agreement	P	Ac		

Table 3 - Responsibility Matrix

Legend:

A = Approve   **Ac** = Accept   **P** = Prepare   **R** = Review

B.3. Managerial process (section 3)

This section defines the type of management being used for the project.

B.3.1. Management objectives & priorities (section 3.1)

This section describes the philosophy, goals, and priorities for management activities during the project. (Section 3.4 contains reporting mechanisms for the project team.)

The following should be specified.

- Reporting frequency and procedures.

- Management priorities with regard to requirements, budget and schedule.

B.3.2. Assumptions, dependencies & constraints (section 3.2)

This section states the assumptions and constraints taken into account in developing the project plan.

For this section the following is an example.

This project is based on the following assumptions.

1. The development hardware for the project will be available by dd/mm/yy.

2. The organisational re-arrangements will be in place prior to the field trial.

B.3.3. Risk management (section 3.3)

Risk Management is used when the project plan is being developed, and at other times in the project life-cycle when new information emerges or significant changes occur whose impact needs to be assessed. Refer to Std-12 - Risk Management.

Risk management tools and techniques are specified in section 4.1.

B.3.3.1. Risk management plan (section 3.3.1)

This section specifies when further iterations of risk management tasks are to be done, and allows for unscheduled risk management to be carried out. Risk Management should be scheduled at the end of one phase and before the next one starts, or when significant new information or change may seriously impact the project.

B.3.3.2. Result of analysis (section 3.3.2)

This section contains the report of the most recent risk management assessment. The information may be provided as a table with references to points following if detail needs to be provided.

For example:

Event or Risk	Probability	Severity	Preventative steps	Contingency
Development hardware late	*Medium*	*High*	*Order on signing of the project agreement.* *Pay a priority fee.*	*Lease suitable hardware.* *Relocate the project.*

Table 4. Risk Analysis table.

B.3.4. Monitoring & controlling mechanisms (section 3.4)

This section shall specify the monitoring and controlling mechanisms to be used for the project.

For the monitoring section, which compares actual versus expected progress, reference may be made to Standard Std-23 - Project Status Reporting for more information.

For the controlling mechanisms, describe the activities to be used in taking remedial action when the monitoring process discovers that there are deviations from the plan.

Control activities may include the following.

- Reworking-plans (but are not confined only to the schedule).

- Implementation of standards.

- Development of procedures.

- Staff training.

- Other control actions may be appropriate.

B.3.5. Staffing plan (section 3.5)

This section contains the most current estimate of numbers and types of personnel required, hence requires regular review and update.

The staffing plan should include the following.

- Required skills.

- Start (availability) times.

- Duration (how long required).

- Recruitment strategy.

- Training plan.

- Phasing out of staff.

It may be necessary to take into consideration the delays which are often experienced when trying to find suitably qualified project staff. The difference between line management and project management becomes more apparent when particular skills are no longer needed on a project and when different skills/numbers may be required for the next phase. Staffing plans need to take into account that some staff are needed for a discrete time only, after which they are no longer needed.

B.4. Technical process (section 4)

This section defines the technical methods, tools and techniques to be used on the project.

B.4.1. Methods, tools & techniques (section 4.1)

To avoid the possibility of confusion, details of methodology, techniques, and tools are specified in the project plan rather than the quality plan. This applies to any methodology, technique, or tool for the support of quality functions.

The quality plan (Section 9) contains a cross-reference to this section.

B.4.1.1. Methodology (section 4.1.1)

This section specifies the methodology to be used for the project, and details any tailoring of the methodology for the current project.

B.4.1.2. Techniques (section 4.1.2)

This refers to information systems management techniques such as the following.

- Project management, planning and control.

- Quality management.

- Testing management.

- Risk management.

- Structured techniques, object oriented techniques (as applicable).

B.4.1.3. Tools (section 4.1.3)

This section specifies the tools used to support the project.

For example.

- **General purpose tools** - word processing, spreadsheet, drawing, scheduling, mail and other office automation tools.

- **Process support tools** - specialist tools including computer aided software engineering, configuration management, quality metrics, risk management.

- **Development and testing tools** - Compilers, development libraries, automated test tools.

B.4.2. Software documentation plan (section 4.2)

This subsection outlines the documentation plan for the project. This includes all project and user documentation. All project documentation shall be prepared in accordance with Std-01.

This section shall specify the following.

1. Software documentation standards, templates and conventions. Reference to Std-01 is sufficient.

2. The software documentation schedule, resources and milestone dates. Strictly this requires a summary of the plans for production of all project, and user

documentation. Rather than include this information directly a reference to the project schedule is sufficient provided that the schedule complies with the following.

- Shows all software documentation related tasks in their proper timing relationships to other project tasks.

- Allows a report to be produced that shows only the software documentation related tasks, and the timing relationships between them.

- Allows a report to be produced that shows all project tasks but highlights the software documentation related tasks.

- *The following example complies with the requirements of this section.*

B.4.3. Project support functions (section 4.3)

B.4.3.1. Quality (section 4.3.1)

The quality plan [Std-08] shall exist as a separate document in its own right and is to be referenced here. The resources and budget needed for project quality assurance shall be included in the quality plan.

B.4.3.2. Testing [verification & validation in IEEE] (section 4.3.2)

This section of the project plan shall specify the following.

1. The testing standards to be used. Any deviations from these standards shall be identified and justified.

2. The decisions concerning which of the following test phases will be performed formally and which will be performed informally.

    - **Module testing**. - all or only specified modules.

    - **Integration testing**. - the number of integration stages.

3. **System testing**.

4. Identify the resources and budget required for testing of the project.

The following example fulfils the requirements of this section.

All testing will be performed in accordance with Std-25 - Testing Process).

In accordance with these standards System testing will be performed formally. Both integration and module testing will be performed informally those test plan documents will be prepared for them. Module and integration testing are being performed informally to meet the tight project time schedule. This can be justified technically as the product is composed of very few modules with very little interaction between them, and excellent visibility of the individual modules from the system boundaries.

The testing related activities account for XX% (should be at least 30-40% or you are in real trouble) of the project budget. This is broken down as follows.

Module testing	20%
Integration testing	10%
System testing	70%

B.4.3.3. Configuration management (section 4.3.3)

The Configuration Management Plan may exist as a separate document in its own right, or for smaller projects, would be provided here. If the configuration management plan is a part of the project plan the resources and budget required for this function shall be stated in this section, otherwise it will be included as part of the separate configuration management plan.

B.4.3.4. Administration/other (section 4.3.4)

Large or special projects may require significant use of support resources not covered in the above sections. (i.e. administrative staff, office automation support, use of a project coordinator or schedular to assist the project manager). Specify these resources here.

B.5. Work packages, schedule & budget (section 5)

This section defines the work break down structure and schedule.

B.5.1. Work packages (section 5.1)

Strictly this section should contain the detailed work breakdown structure for the project, however including such detail in the project plan contributes little to the usefulness of the document. A full work breakdown structure for the project must exist, within the detailed project schedule, as a separate document, or within the methodology documentation. A reference to the location of the full work breakdown structure (normally the project schedule) shall be given, or it shall be attached in an appendix.

This section may give a next layer breakdown of the project life-cycle as discussed in section 2.1 (8.3.2.1) if it is judged that this would be useful to the reader. (This information is also contained within the project schedule, and if given must be kept up to date if the schedule is changed.) For example each project phase (Work package) can be expanded into the next level of detail to provide an overview of the execution of the work.

Note: The term work package is an IEEE term having the same meaning as project phase.

B.5.2. Dependencies (section 5.2)

This section should identify project phases (work packages), or parts thereof, which must be done in a certain sequence or that must overlap in their execution. This information is contained in detail in the project schedule, and duplication of it here contributes little to the usefulness of the project plan. Thus this section should contain a reference to the project schedule.

It may be useful to include dependencies between external events and the project, or between major project activities if this information would be useful to the reader. (This information is also contained within the project schedule, and if given must be kept up to date if the schedule is changed.)

B.5.3. Resource requirements (section 5.3)

This section specifies estimates for the following.

- Skills and numbers of project staff required.

- The development environment required.

- The accommodation, travel and administrative resources required.

B.5.4. Budget & resource allocation (section 5.4)

This section shall give a breakdown of the global project budget and resources to the major project functions. This may include project phases or major activities such as the following.

- Deployment equipment purchases

- Deployment labour costs

- Development costs equipment

- Development costs labour

- Prime contractor costs

- Contract supervision and acceptance testing costs

B.5.5. Schedule (section 5.5)

This section outlines the schedule for the various project functions, activities and tasks, taking into account the precedence relations and the required milestone dates. In effect, it brings together the elements identified in Section 5.1 to 5.4 to provide the schedule for all project functions.

Normally this is a reference to the project schedule in an approved scheduling package. Note that section 8.4.2 of this standard (Project plan section 4.2), and section 8.3.2 of Std-08 (Quality Plan section 3.2) have requirements on the way the schedule is constructed.

At some future time a project schedule standard should be written as part of the software quality management system standards.

B.6. Project training plan (section 6)

This section is only required for projects which justify the need for individual project training, separate to the overall business unit Training Plan.

Note: Implementing overall training and development requirements is the responsibility of the Manager(s) carrying out the systems development strategic development function, and therefore a separate training plan would not normally be necessary for individual projects. However certain projects might need very specific training, requiring a separate training plan, to be determined by the Project Manager. Under such circumstances, the Project Manager shall include a Training Plan as Section 6 of the project plan, as shown in the template.

B.7. Index [Optional] (section 7)

On larger project plans an index aides in accessing information.

B.8. Additional components

Certain additional components may be needed and details of these extras can be appended at the end of the project plan. Additional sections should follow the same numbering system as the rest of the project plan.

Additional components might include some of the following.

- Subcontractor management plans.

- Security plans.

- Independent verification and validation plans.

- Hardware procurement plans.

- Facilities plans.

- Installation plans.

- Data conversion plans.

- System transition plans.

- Product maintenance plans.

www.ingramcontent.com/pod-product-compliance
Lightning Source LLC
Chambersburg PA
CBHW070928050326
40689CB00015B/3665